YOUR KNOWLEDGE HAS VALUE

- We will publish your bachelor's and master's thesis, essays and papers

- Your own eBook and book - sold worldwide in all relevant shops

- Earn money with each sale

Upload your text at www.GRIN.com and publish for free

Cats Tolerate Cold Weather. Most Adult Cats can Go Outdoors in Winter

Kai Aulio

Bibliographic information published by the German National Library:

The German National Library lists this publication in the National Bibliography; detailed bibliographic data are available on the Internet at http://dnb.dnb.de.

ISBN: 9783346789365
This book is also available as an ebook.

Cats Tolerate Cold Weather

Most Adult Cats Can Go Outdoors In Winter

Kai Aulio

December 2022

Content

Abstract

Attitudes towards acceptance of outdoor living of cats (Felis catus) in man's neighborhoods or in wild nature are deeply divided: On the other hand, free roaming is a natural behavioral characteristic, common in all felines, and on the other hand modern pet cats, often bred to satisfy human desires are regarded as unable to withstand harsh environmental conditions, especially cold weather. Most urban pet cats spend all or most of their time indoors, and in the strictest attitudes, these pets do not survive in low temperatures. In literature, the minimum air temperature in cats' living spaces is sometimes set as low as + 25 degrees centigrade (^{o}C). Demands to keep cats indoors are justified by stating the inner (core) body temperature of cats is $38 - 39$ ^{o}C, i.e., a couple of degrees higher than in human beings. Cats have, however, several essential characteristics that make outdoor living possible and safe.

The thick fur, consisting of at least two different hair layers is a very good insulator, allowing the felines to live outdoors even in sub-zero temperatures – as low as - 20 ^{o}C. Besides the warm fur, cats have a well-developed "inner fur", the layers of brown adipose fat tissues. For about a decade ago, the occurrence of brown fat in cats was realized. This tissue is efficient in producing warmth, thus allowing cats to thrive outdoors even in cold weather. And this protecting characteristic is even more enhanced every moment the cat spends in the cold.

The choice of allowing the cats to go outdoors has also a social aspect. Cats living entirely indoors are at greater risk of problematic behavior and poor socialization than felines that can at least occasionally go outdoors. Fearfulness and aggression, and excessive and detrimental grooming are more common in indoor than outdoor cats.

The adaptation to cold weather in cats was first demonstrated in laboratory animals, but nowadays we know the same applies to most adult cats, whether they are used to spending time outdoors. In newborn and very young kittens, old individuals, and sick cats the cold adaptations are not well-developed, and for these felines, man should always secure a warm environment.

Hypothermia, i.e., a marked decrease in core temperature is dangerous, and every cat owner or handler should recognize its signs. Careful warming of an exposed cat is critical because rising the animal's temperature too rapidly can be dangerous, even fatal.

Outdoor felines' natural activity and mobility typically follow a two-peak pattern where the creatures operate mostly in the early morning hours and in the dusk. Outdoor cats rarely move far from home, so they usually can find shelter if the weather conditions suddenly turn harsh.

Introduction

Disputes about whether cats (*Felis catus*) should be allowed to freely roam in nature are eternal questions without the right answers. Considering the history of cats for thousands of years, it is useless to debate, whether a cat can survive outside without help or protection from humans. Cats have lived in close connection with humans at least for 9'000 years. However, the history of cats living in cold climate zones is much shorter (Brown, 2020).

In history, the size, body morphology, and characteristic of skin or feathers covering animal skin have evolved to be optimal in local environmental conditions. Cats living in cold climate zone have several specific characteristics, specially adapted to living in low temperatures. These features ensure that the body in warm-blooded mammals remains stable, both on exceptionally hot summer days and freezing winter nights.

Cats adapted to freezing air temperatures have typically sturdy body morphology, thick legs and round paws, a short tail, a round head, and small ears. Similar structural adaptations are necessary for wild cats living in mountains and highlands, e.g., in Turkey and Iran, where some of the most admired house cats with long and thick fur originate (Fogle, 2003; Brown, 2020).

The first remarks about the coexistence of cats with man were from Ancient Egypt, thousands of years ago. In Africa, the cats probably never had to meet freezing environments. But as early as in the third century C.E. the unparalleled ability of cats to catch small rodents and to control the populations of mice and rats in barns and households united the two species. Pet animals, without obligations to hunt but only keep a company to host families, are historically a rather new phenomenon (Piamore, 2019; Brown, 2020).

Household cats, spending practically their whole life indoors, became common some 50 to 60 years ago. Cats living at least part of the time freely outside, whether in the farmyard and barn or roaming in forests near the owner's house, have existed for centuries, maybe millennia. The cats that provoke sometimes angry reactions and discussions in urban landscapes are not necessarily – and not even often – neglected or abandoned wild creatures, thus being a problem to take care of.

Even though we mostly consider cats as household pets, these felines belong –
and they have always belonged – to nature as an established and natural
component of biota and ecosystems (Brown, 2020; Hill, 2022).

It is obvious that feral and stray cats live practically entirely outdoors – and that
these creatures cope well with even harsh climatic conditions, including cold
weather. For indoor cats, it is the decision of cats' owners, whether the pet can
spend time outdoors, and if yes, can the moggy roam freely or takes the outdoor
activities to place leashed or otherwise under the owners' control. The decision
about the pet cat's living style is entirely in the owners' hands. In this important
choice, several social and demographic characteristics are important – in
addition to the cat's welfare, of course. The factors determining the pet cats'
indoor-outdoor lifestyle were summarized in the exclusive literature review by
British experts in zoology and veterinary sciences (Foreman-Worsley, 2021).

The provision of outdoor access is dependent on geography. In the United
States and Canada, most pet cats are entirely tied to indoor living, whereas in
Europe, Australia, and New Zealand the cats' possibility to go outdoors – free
or in owners' control – is prevailing. The owners' demographics are also
important. According to the international survey, the age of the owners of "the
indoor only category" is 26–35 years, and these households have often more
than one cat. The cats that have higher odds of outdoor access live typically in
households of people in their middle age or senior citizens, and in these
households, the cats are often mature and most often male (Foreman-Worsley,
2021).

For the owners of outdoor cats, the main reason to allow free roaming or
controlled outdoor activities is the well-being of the pets. Possibilities to follow
species-specific behavior is considered necessary for the cats' mental and
physical health.

The reasons the owner decides to keep cats entirely indoors are varied. The
most important factor is security, especially aiming to protect cats from traffic
accidents. In North America, the second factor is protecting pets from predation
by wild animals. In Europe and Australia, the second reason for restricting
outdoor access is protection from hunters' and catheters' vandalism (Foreman-
Worsley, 2021).

Permitting the cats' free-roaming outdoors is a controversial question – with no
right answers. The most often presented argument against outdoor cats is

hunting, especially the fact that cats kill millions or even billions of birds and other small animals. The undeniable fact is that cats are natural-born killers. But the number of prey the cats take annually is often overestimated. A crude estimate is that in the prey assortment birds represent one-third, and other small animals such as rodents cover two-thirds of the annually killed natural creatures.

Predation of birds and other small animals by free-roaming cats is only a part of the total consequence of free free-roaming cats. Besides the kills, cats can spread diseases. Studies and follow-ups of cats' effects are concentrated on urban and conservation areas in developed countries, and most of the total influences in other parts of the earth are understudied (Loss *et al.*, 2022).

Nobody hardly denies the positive effect that hunting cats has in keeping the populations of detrimental rats and mice in balance in farms and other residential areas (Fogle, 2003). The effects of cats' predation take place principally very near the animals' residences, and so the actual number of kills is significant only in restricted areas. Despite the huge number of prey animals, the influence on nature's biodiversity is negligible. Only in certain conservation areas, cats can cause a fatal reduction of rare or endangered animal taxa (Baker *et al.*, 2010; Kays et al., 2020).

Indoor cat: new breed for urban society

The number of cats living with people is steeply increased, and the exceptional era of the COVID-19 pandemic further stimulated the acquisition of pets. A rather new feature in the relationship between humans and animals is the tendency to have cats that only live indoors. As a matter of fact, indoor cats are often considered – without any scientific basis or justification – as a special feline breed or even a species.

The dominance of indoor living habits is emphasized by textbooks and pet caretaker manuals that are written entirely for indoor cats (Bessant, 2006). But staying outdoors should rather be a rule than an exception for *Felis catus*. The detention of a pet entirely indoors can hardly be considered a species-specific feature for a cat.

Manuals for indoor cat owners are naturally necessary, especially for new cat households. Walking a cat on a leash and/or fencing the garden and yards to enable at least a vision of outdoor living are recommended and will certainly improve the welfare of any cat.

As the renowned cat expert's textbook summarizes: "Cats need to behave like cats" (Halls, 2010). That is, restricting the cat's possibilities to free mobility and exercise leads to boredom and frustration, and eventually risks both the physical and emotional well-being of the cat. If it is impossible to provide outdoor excess, the indoor environment must be constructed as diverse as possible, e.g., by providing climbing trees, terraces, and hiding places, mimicking the natural habitats.

The discussion about indoor vs. outdoor cats is often black and white, but there are intermediate forms of having and caring for felines (Heath, 2003). To secure the cat's well-being the owner must recognize and respect the background of the animal. If a cat is used to living outside, keeping such a feline entirely indoors can cause serious mental and behavioral problems. When the owner takes the pet as a newborn kitten, the habituation to an indoor-only lifestyle usually succeeds well (Heath, 2003).

Restricting mobility can be a health risk

The choice of keeping a cat entirely indoors or enabling the pet to roam outdoors determines several social/antisocial features in the feline's life. In the comprehensive study of 3'200 cats' behavioral features, the study by an expert group in Finland found a strong connection between problematic behavior and poor socialization in indoor cats. The group led by Hannes Lohi, a professor in Veterinary Molecular Genetics at the University of Helsinki used some thirty features in cats' behavioral, and environmental characteristics in evaluating the connections between fearfulness and social behaviors in cats (Mikkola *et al.*, 2022).

In the rich mixture of social characteristics studied, the Finnish study identified a clear connection between indoor/outdoor lifestyle and socialization in cats. The possibility to stay outdoors, at least to a limited extent in the owner's balcony or garden enhanced the quality of the pets' behavior. Pet cats that had the possibility of freely roaming outside had markedly less problematic behavior than cats that lived entirely indoors. The possibility of living outdoors diminished aggression and detrimental excessive grooming of the creature's fur (Mikkola *et al.*, 2022).

In the United States, the comprehensive review by the Purdue University of factors affecting indoor cats' welfare showed that limited space available provokes aggression and/or anxiety. Therefore, cats kept primarily indoors and maintained without access to the outdoors have an increased risk of relinquishment (Stella & Croney, 2016). The study of the welfare of 1'177 cats from 550 owners in Germany showed that the possibility of outdoor access decreased the cats' negative behavior, concluding "Few or no problems with cats that could go outside" (Heidenberger, 1997).

Roaming around, exploring new territories, and hunting are natural phenomena for all cats, and suppressing these functions can cause distress to them. Keeping cats too tightly chained can lead to stress-related diseases (Milne & Wild, 2018).

Pros and cons of cats' outdoor living

Public debate on the cats' justification for roaming freely outdoors, especially in the wild, is primarily concentrated on the several negative effects. These phenomena include health effects for the cat itself, i.e., the increased possibility for infections and injuries caused by other animals and especially traffic. The list of unpleasant consequences includes the risk of swallowing toxic substances. A negative aspect often linked to outdoor cats is the harm the creatures cause in the neighborhood.

Most complaints arise from defecation in other people's gardens or yards, and especially in the breeding season, the noisy caterwauling by the male cats is experienced as annoying. From the owners' point of view, the elevated risk of losing the pet when a roamer is lost or stolen is a considerable subject (Tan *et al.*, 2020).

In the temperate and cold climate zones, outdoor cats face a special danger: Exposure to antifreeze chemicals. The substance used in car radiators and some cleaning agents is typically ethylene glycol. This chemical is broken down in the liver, and the resulting molecules are very toxic, causing severe illness and in the worst-case scenario, the death of a cat licking antifreeze spilled to the ground (International Cat Care, 2020).

On the other hand, the possibility of outdoor living enhances the cat's welfare in many ways. The possibility of free outdoor roaming is an important factor in meeting the cats' behavioral needs Climbing upwards to observe the environment, and above all the possibility of hunting are natural features of all cats, and rendering these activities is difficult for indoor-only cats. Discussions often ignore that cats – all cats regardless of the owner's desires – are natural-born hunters and killers. So, the possibility of stalking and hunting small animals is essential to any cat's welfare (Tan *et al.*, 2020). The physical exercise provided by outdoor roaming keeps the cats physically fit. Indoor cats are usually more passive, leading to obesity which is both aesthetically unpleasant and increases the risk of several diseases.

Outdoor cats may need help but mostly they cope on their own

The fact that there are millions of cats living outside in cold environments accompanied by man has led to numerous practical manuals to take care of felines. Whether or not the cats need some assistance, a few actions are at all cases welcome and necessary for outdoor cats. Shelter, food and water, and security are the main points on the agenda of cat protection (Tractive, 2022).

Supply of unfrozen food and water is essential, even though cats are predators that should be able to get the food on their own. In winter, the hunting and scavenging possibilities are scarce, and so the food service for owned as much as for feral and stray cats is essential. In services, it is important to secure that the supply keeps unfrozen.

Cats are naturally very clever in finding shelter. Lithe bodies slip gracefully inside through even the smallest cracks but in open yards, exclusive cat housing is valuable, often a lifesaver. The shelter can be a simple cardboard box, lined with straw or other material that keeps dry and unfrozen even in sub-zero temperatures (Cat Protection, 2022; Wagner *et el.*, 2018).

Of several life-threatening dangers for outdoor, free-roaming cats the most often fatal ones are traffic accidents, especially in urban neighborhoods. Chemical hazards are the other danger that always threatens cats. Of the toxic substances that cats often face in their outdoor environments are rodenticides, i.e., poisons that kill noxious rats and mice. Hunting small animals weakened by poisons is easy prey for cats, and scavenging dead animals is even more dangerous. In cold environments, the possibility of consuming toxic antifreeze chemicals, used in car radiators and non-freezing washing liquids can lead to fatal symptoms in cats (International Cat Care, 2020).

Anger and complaints caused by the fact that cats kill birds and other small animals have strengthened as the number of pets has increased in recent years or decades. This fanaticism ignores the fact that cats are carnivores, i.e., natural-born killers. All the cats are hunters, even the ones that are most tightly bound living entirely inside – and the most beautiful pedigree cats that are winners in shows. Domestic cats have not lost their instincts as predators: Hunting is in the genome of cats, and predation has nothing to do with hunger. Prowling, running, and attacking the prey are basic needs that every cat has in its genes. The possibility to follow natural needs is only up to the cat's owner (Fogle, 2003).

All outdoor cats need food, water, and shelter

Meeting a cat outdoors in cold, freezing weather evokes sympathy, and possibly the desire to help the poor creature. If so, the first thing to do is to try to find the owner of the cat. If lucky, the roamer wears a collar with address details. In many countries, cats have a mandatory microchip, so with help from authorities or charity organizations, the home is certainly found.

When meeting a cat, no one can be sure if the creature needs help at all. House cats often roam around near their residences, and they can cope with short periods even in harsh weather. Passing a cat with some food can in most cases tell the "quality" of a cat. House cats are social, and they approach humans – at least when provided with food. Feral cats, on the other hand, belong to someone, and they are just occasionally searching for adventure. They can be approached, and the same kind of help can be provided as for the house cats.

Not to mention wildcats. They have never lived with human care or contacts, and they really do not need any help – and almost certainly they do not even accept help. The actual wildcats are exceptionally clever in finding shelter during cold weather or other harsh conditions. General guidance in meeting a proper wildcat in nature is: A wildcat prefers living freely, alone or with their peers, over the possibility to move inside with people – even during winter's freezing weather (Animal Humane Society, 2022).

In outdoor feeding, it is extremely important – and difficult at the same time – to ensure that the food and water keep unfrozen. Special feeding bowls and frequent drinking water changes are a lifeline for the welfare of outdoor cats. In cold environments, the need for energy will inevitably increase, so the cat must be able to get food more often than just a normal feeding schedule. Detailed instructions in these regards are outside the scope of the present review, but extensive literature is available in print or online (Animal Humane Society, 2022; Cat Protection, 2022; Tractive, 2022).

Cold tolerance varies between breeds

Cats have conquered practically all the continents, and that's why the animals must withstand all kinds of natural and weather conditions – even the exposition to very cold air occasionally. Living in freezing conditions is harsh, and that's why cats have evolved into the extremely adaptable species the felines are today. There are, however, cats with many faces: The moggies living entirely inside in warm houses do not need resistance to cold weather, and so they are not naturally prepared to meet freezing conditions. The cats living permanently or at least occasionally freely outside are better equipped to withstand the freeze. Numerous fanatic discussions in online chats call for a total ban on keeping house cats freely outside during winter months in temperate or cold environments.

When discussing the welfare of cats in cold environments one must recognize that there are many kinds of cats – in addition to the dual classification between indoor cats and outdoor cats. These two cats are the same – the only difference is created by man – the choice made by the owner or holder of the cat. Urban cats are often indoor creatures, spending their whole life in homes as pets and as a pleasure for humans. In the countryside, cats have an actual role as a working force while predating mice and rats (Argos, 2022).

In any evaluation of the ability of cats to withstand cold or freezing environments, it is important to consider the breed of the feline. Several breeds are well adapted to cold, but for certain cat breeds, even mild temperatures are biologically uninhabitable. The cats with long hair and dense fur can withstand cold, even without any help provided by a man. The Norwegian Forest Cats and Persian cats are examples of breeds with suitable characteristics for living in cold (Brown, 2020; Argos, 2022).

Healthy adults tolerate freezing temperatures

If a cat, any cat, is in good condition, the animal can spend time outdoors, even when the air temperature falls below zero degrees centigrade (0 °C), providing that the cat is healthy and adult. Old and very young cats – and of course sick individuals – should be kept inside and safely in warm environments. Most adult cats can spend even long periods of time outside in freezing temperatures, even as low as - 20 °C. This applies also to spending nights outside during winter months – naturally provided that the cat has the possibility to find shelter if necessary. Young kittens, with an age of fewer than six months, should not be outside in cold weather and snowy conditions because the inner woolly fur has not yet properly developed (Ein Herz für Tiere, 2020).

Special attention to the temperatures in pet cats' living environments must be given to very young kittens. A newborn kitten can suffer from hypothermia in temperatures that are quite normal for most cats. The temperature regulation in kittens and the ability of the cat to maintain or increase the body temperature are undeveloped in young ones (Pet Med, 2009).

The ability to actively regulate and/or maintain the inner body temperature is well-developed in adult cats but kittens are dependent on the warmth of their environment. A newborn cat is unable to regulate body temperature during the first three weeks of their lives. The small kittens are practically bound to the temperature and shelter provided by touch and nearness by the parents and siblings. In cats, active temperature regulation is developed at the age of about six to seven weeks (Brown, 2020).

The schedule in the development of the cats' temperature regulation was verified in the experimental studies at the University of California in the 1970s. During the first week, the body temperatures of newborn kittens were +37.0 °C on average, i.e., very low on the scale typical for cats. In about seven weeks, the body temperature rose to the typical levels for house cats, + 38.2 °C (Olmstedt *et al.*, 1979).

The body temperature in very young kittens reacts very rapidly – and possibly dangerously – to the cooling of the cat's environment. When kittens with an age of fewer than two weeks were kept at a room temperature of +25 °C, the inner body temperature of the animal decreased in the schedule of 0.02 °C per minute.

New-born cats react to marked changes in outside temperatures, whether the changes are warm or cold. As kittens were transferred to plus 60 °C or minus 15 °C, they tried immediately and actively moving away from those extreme environments.

When exposed to the temperature of minus 15 degrees centigrade, the inner body temperature of a one-week-old kitten decreased by 0.2 °C per minute. The reaction was not restricted to newborns. At the age of six weeks, the decrease in body temperature was remarkable, at the pace of 0.1 °C per minute.

As opposed to kittens, adult cats can actively regulate and maintain body temperatures using shivering (activating muscles), constructing veins, and enhancing the insulation (increasing the amount of air between the hair) when bristling their fur (Olmstedt *et al.*, 1979).

Important to recognize cats' needs

Any owner or keeper of a pet cat should recognize situations where the cat yarns or even absolutely requires extra warmth. In a cold environment, a cat starts shivering, and when the animal feels freezing, the whole body starts trembling. When in cold, the muscle activity starts producing warmth, but this temporary help comes with a price: Active muscle work needs energy, and therefore cats must eat more than usual.

In addition to cats' movements, the moisture of the fur is very important in temperature maintenance. The cat fur is an excellent repellent – but only when the hairs are dry. When moist or wet, the insulation capacity rapidly decreases or disappears. Freezing of wet fur in air temperatures below zero degrees centigrade can easily cause the death of a cat.

In severe frost and during long-time exposure to a cold environment, risks in cats are just the same as with humans: A cat can get frostbite in external body organs, such as tips of ears, tail, paws, toes, and nose (Medivet, 2022).

Dangers with hypothermia

Every cat owner should recognize the symptoms of hypothermia, i.e., the situation where the body temperature of an animal has fallen below normal. In cats, this is challenging because the temperature regime and optimal degrees are higher than in humans. In any case of even mild hypothermia, it is essential to provide extra warmth, the right kind of food, and drinking water as soon as possible. Making the cold-exposed cat eat is essential because energy consumption and thus the need for extra energy tend to increase due to the decrease in body temperature.

The good news with hypothermia is that the condition – though seemingly dramatic at first – is usually cured quite easily and safely. Positive prognosis is common even in extreme situations, even when the cat was momentarily in a coma or suffering from cardiac arrest (Broderus *et al.*, 2017).

The external signs and negative changes in metabolism in hypothermia are the results of very dangerous threats to cats' well-being and health. With every degree's decrease in inner body temperature, the efficiency of cerebral circulation weakens by 6–7 percent. Such a marked change results in severe deterioration in both physical and mental capacity (Dozeman & Dacvecc, 2020).

Cat's body temperature is higher than in men

The body temperature of a cat is 1–2 °C higher than the temperature of people – even though both animal species are classified as warm-blooded mammals. The normal body temperature in cats– as measured by the most accurate method, i.e., rectal temperature – lies between 38.2 °C – 39.2 °C (PetMed, 2009).

Hypothermia is the decrease in normal body temperature that results from external factors, i.e., long-term exposure to cold air, water, or snow. In human beings, the extent of hypothermia is usually classified as life-threatening when the inner (core) temperature falls to 35 °C or lower (Peiris *et al.*, 2018). The characteristics, dangers, and procedures of severe hypothermia determined for man apply in most cases in all warm-blooded mammals. In the case of felines, it is important to recognize that the cats' core temperature is at least two degrees higher than that of man.

In the medical and veterinary literature, the extent of hypothermia in cats is usually classified into three categories (PetMed, 2009):

● Mild hypothermia, with body temperature between 32 °C – 35 °C,
● Average hypothermia, with body temperature between 28 °C – 32 °C,
● Severe hypothermia, with body temperature less than 28 °C

In the United States, the veterinary literature usually determines the threshold temperature for severe hypothermia to be any level below 30 °C (Dozeman & Dacvecc, 2020).

Irreversible, i.e., state that cannot be restored to normal, and the level where a cat will probably die, is 24 °C. There are, however, several reports in the veterinary literature, where a cat has survived even when the inner body temperature has fallen as low as 16 °C to 20 °C. Long-term hypothermia is dangerous, and usually fatal to all species of warm-blooded animals, both in human beings and in cats.

In warm-blooded mammals, the body temperature tends to maintain stability, but occasional variations are typical due to changing activity in vital functions and physical activity. The time of the day or night determines the body temperature. In human beings, the body temperature is at its lowest level early in the morning, and peak temperatures occur in the late afternoon hours. In cats, on the opposite, the cycle of temperature variation is different. The core temperature in felines is higher than in men, and both minimum and maximum degrees are achieved at different hours of the day (Dozeman & Dacvecc, 2020).

Outdoor living enhances cold tolerance

The common view of cats as creatures that only thrive in warm environments is rather new. The phenomenon called *indoor cats* was born alongside urbanization and pets living indoors. On the other hand, anyone practicing even minor exercise or any outdoor hobbies, e.g., walking a dog, can observe cats dashing around, even during the winter's cold, freezing weather. Most of those cats are not pets abandoned by their owners or suffering from exposure to outdoor living.

The instinct of cats to hunt draws even the most experienced indoor cat to head towards fresh air – regardless of the quality and quantity of household food and accommodation services. A unique story is about barn cats living on farms. The purpose of life in these predators is to hunt rodents, mice, and rats, thus being priceless to their owners. Barn cats are used to spending most of their time outdoors – or at least away from the owners' residence – throughout the year – in winter months, too. These cats can look for and find shelter independently, without any help or guidance from their owners. Barn cats always find warm shelter if the weather turns out to be too freezing.

Cats have several means to cope in cold environments. Adaptation to cool and cold weather or frost is a species-specific characteristic for cats – all cats, regardless of the breed. Urban cats, often entirely indoor pets, do not need an adaptation to withstand cold. But in the genome, every cat has valuable assets to cope with even in harsh conditions. The well-developed feature of cats' cold hardiness has been well-studied for decades (Brown, 2020).

In the experimental, long-term study at the University of Marburg in Germany, the researchers kept two groups of cats either warmer than the households' conventional indoor temperature or in very cold, aerated chambers under artificial light. The cats spent two years either at + 5 °C or at + 30 °C. After 24 months the groups changed their places, and in the following 36 months the cats previously living in warm conditions moved to cold chambers, and the cold-adapted cats moved to warm rooms. After the second phase of the long study period, several characteristics typical for thermoregulation and adaptation to varying living temperatures were analyzed (Hensel & Banet, 1982).

The most important finding in the physiological outcome was the enhancement of metabolism by 45 percent in cats adapted to cold living conditions. These cats were able to compensate for the increased energy requirements by eating more food than the cats in the reference group.

The cats adapted to cold environments rather rapidly and got the ability to react to changing conditions. In this group, the onset of the necessary reaction to change started at eight degrees lower temperature than in cats that were not adapted to cold.

The long-term cold adaptation changed the capacity of the cats to survive in freezing conditions. When exposed to a temperature of five degrees below zero (- 5 °C), the insulation capacity of cats' fur weakened in both groups of cats. The insulation rate in the cats transferred to freezing conditions was enhanced by 34 percent in cats adapted to warm conditions. On the other hand, in the group that had spent time in the cold, no change in the fur was realized. The insulation of fur in these cats was good enough without any further enhancement (Hensel & Banet, 1982).

The adaptation process is always a "personal one", i.e., each animal reacts differently to changing or varying environments. When moving to a new address and territory, the change inevitably influences the cat's behavior and well-being. The amplitude of the influence varies according to the magnitude of change.

If the range of temperature extremes and seasonal changes are close to each other, the adaptation process is easy, and the cat can feel at home a week after settling into the new address. When the range is wide, for example when moving from a hot climate zone into a cold northern zone, the adaption process can take as long as a year (AVMA, 2022).

Domestic cats stay near home

The territory – the area where a cat principally lives and hunts – in house cats and urban feral cats is remarkably small. Cats living with or at least in close connection with humans occupy areas of only a few hectares, whereas actual wildcats roam around in territories covering 70 hectares or more. Of the cats with a close relationship with man, free-living barn cats in the countryside govern and use for hunting an area of up to 60 hectares (Morris, 1994).

In an international study from six countries, led by Roland Kays at the North Carolina Museum of Natural Sciences USA, the average size of a cat's territory was 3.6 hectares, and only a few exceptional individuals occupied a territory wider than one square kilometer (1 km^2). The size of a cat's territory was not related to the presence or absence of other predators, so this living area is not determined by competition between species (Kays *et al.*, 2020).

Three-quarters of the cats studied lived entirely in disturbed areas with significant anthropogenic influences. Thus, the cats' role in natural habitats with high biodiversity seems to be minuscule.

Cats are natural predators and so they affect the numbers and occasionally the species relationships of small animals within their territories. Due to the small size of a territory, the effect of any cat is only slight. A pet cat's effect on biota is concentrated within 100 meters of their homes (Kays *et al.*, 2020).

The idea of very limited territory outdoor cats roam was further demonstrated by the study by Roland Kays and Amielle DeWan. In their study of adult indoor/outdoor cats, i.e., pet cats that live in the owners' houses but are at least part-time free to move around freely in the neighborhood. The area these cats used was limited to the immediate vicinity of the residence – the garden or yard of their own or their neighbors' properties. Very seldom the cats moved into the nearby forest, despite the diverse biota in this nature reserve (Kays & DeWann, 2004).

The number of prey the hunting cats brought home was smaller than the average for feral cats, and practically all the animals killed were hunted in open space or within a maximum of 10 meters from the forest edge. One conclusion the authors made was that cold weather limited the cats' outdoor activity and thus the number of prey. The narrow space the cats used allowed the pets to move into a shelter whenever the cold or otherwise unpleasant conditions prevailed (Kays & DeWann, 2004).

In the study led by Zhenwei Zhang at Nanjing University, Chinese and Pakistani researchers determined the home ranges of free-living cats in China. In the follow-up study of adult male and female cats, the territories were even more narrow than the areas presented by the multinational study by Kays *et al.*

(2020). Cats were followed by GPS units for nearly two years in an urban environment in a university campus area.

The maximum home range of 19.8 hectares was determined in adult males in the breeding season. In male cats, the average roaming area was 12.6 ha. On the opposite scale, the smallest home range of a cat covered only 0.6 ha. In females, the home ranges were significantly smaller (5.0–7.7 ha), and more even throughout the year (Zhang *et al.*, 2022).

Resistance to cold is an individual feature, and no specific temperature limit cannot be given for the time or frequency of a house cat's outdoor living. The view often presented in literature and online discussions that cats can live only indoors emphasizes the optima of cats and the fondness of most pet cats. The data presented both experimentally and by empirical experience show that any healthy cat can easily tolerate cold weather, at least as low as minus 20 degrees centigrade (- 20 °C). To do so a cat must be an adult – and this is important – a healthy adult with well-conditioned fur! (Nadine, 2020).

Cats are prepared to withstand the cold

It is common knowledge that the fur gets thicker when the cat lives outdoors throughout the year. Outdoor cats have thick winter fur, and as a characteristic feature their inner, woolly fur layer gets stronger due to exposure to cold. Thick fur, a common feature in the house and barn cats, is a necessity for the forest cats (*Felis silvestris*), so-called wildcats, living in Central Europe.

The thermal insulation provided by the cat's fur is based on the increased volume of air between individual hairs. The more air there is between solitary hairs, the better the fur prevents cold outside air to reach the animal's skin. The same can be seen in birds ruffling their feathers n freezing weather (Brown, 2020; Cats Protection, 2022).

The thermal insulation – against the heat in summer and against cold in winter – is a characteristic varying based on the cats' needs. In healthy cats, the thin summer fur is changed to a thicker and fluffy winter coat in autumn, with the gradually chilling temperatures. The role of base fur is central in enhancing cold adaptation. To materialize, "the cat must know that a change is coming". The cats living entirely indoors in even room temperatures and in artificial lights may not be prepared for the needs, and thus the thickening of fur is weaker than in outdoor cats – and so these pets are not maximally prepared to withstand the cold of winter.

Well-developed heat insulation is seen only in adult cats that spend regular time outside. Very young and very old – as well as sick or infirm – cats have only limited resistance to thermal extremes. Thus, it should be obvious to every cat owner or holder, that outdoor life is regulated strictly based on the individual condition of the pet.

Due to the natural adaptation mechanisms, most cats can withstand cold air temperatures and spend time outdoors in freezing temperatures as low as minus 20 degrees centigrade (- 20 ºC). There are, however, bald, and short-haired cat breeds that have no protective shield against cold weather, and they should not be exposed to harsh winter weather (Uelzener Magazin, 2018; Cats Protection, 2022; Lieblingstier, 2022).

The advice quoted above was given by German and British experts and authorities – i.e., data was originating from a temperate climate zone, where extremely cold weathers are rare. In any case, the conclusions about cats' cold tolerance are right. The fact that cats can tolerate freezing temperatures and spend time outdoors in winter is confirmed by experts from the ultimate North by Swedish vets (Veterinären nu, 2022) and by an insurance company specializing in animal welfare (Svedea, 2022).

Are cats even better equipped than dogs?

Owing to the characteristics of the fur, and the active and thorough maintenance efforts by the fur's owner, cats may be better equipped against wintery cold than dogs. And according to common knowledge, dogs can spend time outdoors even though the thermometer showed minus degrees. In an average cat's dorsal (the backside) fur there are some 10'000 solitary hairs per square centimeter. This is a high number, but on the abdominal side, the cat has more than 20'000 hairs per square centimeter. As a comparison, an average dog has "only" 10'000 hairs per square centimeter (Coop Zeitung, 2011).

There are three main types of hair in a cat's fur (Brown, 2020). Primary guard hairs of the uppermost layer maintain the skin dry and provide air spaces that increase thermal insulation. The undercoat comprising of short and soft hair serves mainly to secure warmth. In addition, the base fur comprises bristly awn hairs.

The cats do deserve good thermal regulation abilities because the animals are diligent to maintain these properties. Often repeated licking of fur enhances the fur's ability to repel water and moisture. Water is, indeed, more dangerous to the cats' cold resistance than the actual degrees of cold or warmth. When moist, the cat's fur weakens significantly, and thus the hairs do not prevent the cold air from entering straight into the skin.

Temperature peaks at night, regardless of activity

The activity of free outdoor cats varies considerably during the 24 hours daily cycle. Typically, the activity of cats is at the highest level during the night, when they move more than in other periods. In the night-time hours, they also hunt more than in the daylight. The relationship between the cats' activity and the animals' body temperature was studied by Australian and German researchers. In the study, eight cats were followed using attached sensors and GPS follow-up continuously (24/7) for two weeks (Hilmer *et al.*, 2010b).

The timing of analysis was the most important factor in determining the variations in the cat's body temperature. The temperature was at its highest level during the night, i.e., the time when the cats moved the most.

In the detailed analysis of the follow-up data, there was a marked correlation between the time and length of daily travel and the body temperature during the daytime hours. At night, when the feline body temperature is naturally at its highest peak, the inner temperature was practically constant, regardless of the moving activity.

As a conclusion of the large-scale follow-up, the research led by Stefanie Hilmer showed that the body temperature in cats is principally determined by genetic, species-specific characteristics and that the nocturnal body temperature rise is not dependent on the cats' activity during the night (Hilmer *et al.,* 2010b).

In another study, the researchers at Iowa University registered a steady, two-peak temperature rhythm. The regular variation in the cats' temperature remains stable irrespective of environmental (air) temperatures or the variation in the cats' physical activity (Randall *et al.*, 1987). The living conditions, i.e., the size of the residence and the cats' possibilities to free outdoor access into a garden or yard markedly influence the pet cats' daily behavior and activity. In the study led by Giuseppe Piccione at the University of Messina, Italy, the cats' activity appeared to adapt to the rhythm of their owners (Piccione *et al.*, 2013).

The indoor cats adapted closely to the daytime activity, resembling the everyday routines of their owners. This is an exception to the two-peak "natural rhythm" observed in cats that can choose the sites and times of activities.

Bimodal daily rhythm – previously found in laboratory animals' behavior – in locomotive and feeding activity was also proved in comparison in the experimental study led by Marine Parker at the University of Strasbourg. The comparison between indoor (in 29 square meter residence) and outdoor (in 1'145 square meter outdoor enclosure) showed surprisingly similar activities in the two groups, with peaking activities in the dim light hours both in the early and late hours of the day (Parker *et al.*, 2022).

Wildcats are warmer than house cats

In feral cats, i.e., felines living freely in nature, independent of any relationship to humans, the body temperature follows a two-peaked daily rhythm. Two-peaked rhythm is typical for house cats, too, but variations differ in both temperature range and temporal pace. The inner body temperature in wild cats is higher than in house cats, and the daily variations are higher in freely living felines.

Daily variations and rhythms of temperature in wildcats were studied by the Australian Department of Environment and Conservation and German researchers. In the first phase of the study, the body temperatures of wildcats were followed in natural conditions, i.e., while the felines followed their independent routines, and in the second phase, the same cats were analyzed after they had spent one year in human custody (Hilmer *et al.*, 2010a).

When living freely and on their own terms in nature, the maximum daily temperature of 39.2 °C was recorded at night, corresponding to the normal, previously determined natural peak. In the daytime, i.e., the period cats are normally resting, the average temperature of the felines was one degree lower (38.1 °C, on average).

The lifestyle of the cats is reflected in the variation of felines' body temperature. In cats living freely in nature, the daily temperature maxima were recorded late in the evening, near the midnight hour (range 22:34 – 23:17). On the contrary, while being in human custody, the cats changed their way of life, and the former nocturnal creatures turned out to be day active. At the same time, the maximum body temperature was measured at noon or early afternoon (range 12:00 – 16:25), instead of midnight.

In addition to the variation in temporal rhythm, the variations in body temperature changed markedly as the wildcats became indoor cats. When living freely outdoors the range in the cats' body temperature was 35.5 °C – 41.9 °C, whereas while living in controlled human care, the body temperature remained rather constant, varying between 36 °C – 39 °C (Hilmer *et al.*, 2010a).

Cold experience enables living outdoors, even in winter

Cats can adapt their living in ways that render their survival and welfare even in very cold outdoor environments. The key to achieving permanent cold resistance is long-term acclimatization, showed a study by Thomas Adams with the US Space Agency in the 1960s. In the experiments, the cats were kept for more than two months either at the temperature of + 25 °C or at + 5 °C. The body temperature and metabolic activity of the two groups were analyzed in two hours periods at + 23 °C, + 10 °C, and 0 °C temperatures (Adams, 1963).

In cats kept in the cold, the adaptation to external conditions was determined by registering an increase in the body's surface temperature. The group kept at zero degrees' temperature showed the adaptation by increasing the movements of the veins through contraction and extension.

The actual and permanent cold resistance was confirmed by the cats that after spending the accommodation period at five degrees had higher body temperatures during the 10 °C and 0 °C experiments (Adams, 1963).

The acclimatization period of two months proved to be an efficient means to increase the cats' possibilities to live outdoors at temperatures, "much colder than most textbooks claim".

The body resists hypothermia, but rewarming can be fatal

A decrease in body temperature below the normal level – hypothermia – is very dangerous if the condition lasts for a long period. But fortunately, cats usually cope with short-term hypothermia without any harm.

The effects of hypothermia have been extensively studied in veterinary sciences. An experimental study analyzed dangerous/fatal damages to circulatory organs in cats that were kept for 48 hours at a temperature of + 29 °C. Only results obtained with control felines are described here. The procedures and results of experiments where damages were intentionally caused and the outcome together with hypothermia are out of the scope of this publication (Steen *et al.*, 1979).

The data viable here are from experiments with cats and monkeys. The study, led by doctor Pettan Steen at the Mayo Medical Scholl and Mayo Clinic studied the effects of artificial hypothermia on the survival and reactions of the animals during the surgery and after the rewarming after operations. Intentional induction of hypothermia is a normal procedure in surgical operations in warm-blooded animals, including man.

All the cats tolerated well the hypothermia of + 29 °C for two days (48 hours). The animals were followed and surveyed continuously via procedures used in intensive care units of hospitals. After the period of hypothermia, the body temperature of the cats was gradually increased by using a thermal blanket and external heaters. The rewarming period turned out to be the most critical phase in surgical procedures.

The experiments with the Mayo Clinic proved the common knowledge that cats are sensitive organisms as regards to temperature. All the cats that had the surgery and spent two days under hypothermia survived the operation, but unfortunately, every cat died during the rewarming to achieve a normal cat body temperature (Steen *et al.*, 1979).

Brown fat is important in cold resistance

Of the several tissue types in humans, the discovery of brown fat (Brown Adipose Fat; BAT) prompted scientific interest quite recently, although the first data of its occurrence were recorded more than a century ago. Special interest in brown fat has gained, however, in recent years, as data on the tissue's occurrences and properties has deepened. Brown fat got a special interest as a "tissue of modern man". The role of this tissue type in the regulation of diabetes and overweight and weight control is of special interest (Cannon & Nedergaard, 2004).

In addition to weight and sugar metabolism, brown fat has proved to be an important factor in temperature regulation in warm-blooded animals. The role of brown fat is important in cold environments. In the cold, the role of brown fat as energy storage turns out to be a producer of energy for mammalian metabolism.

The important role of brown fat in energy metabolism is based on the vast quantity of mitochondria, and a great amount of small fat droplets, as well as the abundance of blood vessels. As soon as the brown fat tissues get activated, the cells of the tissue start producing energy from the stored sugars and amino acids.

On the cellular level, the energy metabolism processes take place on the cell membranes due to the activation of key enzymes. In the process, the synthesis of ATP – the energy-producing machinery of living cells – means that the energy storage stops, and at the same time, energy production at the cellular level gets started.

In warm-blooded animals, the first – and the simplest – step in cold acclimation is to produce warmth by the means of physical activity. This is an automatic process when the subject starts shivering with cold. At the same time, this starts an independent warmth production, independent of muscle work. Brown fat is a key factor in this latter process when fat cells switch off the energy storage and direct the metabolism towards reactions to liberate energy.

In animals exposed to cold the amount and activity of brown fat increase. In warm-blooded animals that have brown fat, this tissue is the most important means of warming processes besides active muscle activity (Cannon & Nedergaard, 2004).

Brown fat in cats was found in 2013

The occurrence of brown fat in humans was scientifically proven as recently as 2009 but nowadays this tissue is the focus of research all over the world. In the cold climate of the North, the presence of brown fat is very important for any animal species, as chronic cold exposure is the most important and efficient means to activate this tissue. The presence of brown fat is now shown in several species of warm-blooded animal species. In 2013 research by the Veterinary College of the University of Illinois in the US described the existence of brown fat in cats (Clark *et al.*, 2013).

The finding came out in research with 11 adult cats (8 to 12 years old) with varying body weights. Of the study subjects, six cats had normal weight, and five cats were overweight. The presence of brown fat was confirmed by the uncoupling protein 1 (UPC1), regulating the metabolism of fat cells.

The presence and activity of the key enzyme were shown in the subcutaneous tissues (tissue straight beneath the skin surface) in all the cats studied. In addition, brown fat was found in tissues surrounding kidneys in 10 of the eleven cats (Clark *et al.*, 2013).

The finding of brown fat should not have been such a surprise because tissues with similar properties were recognized in young kittens decades ago. But it was the 1989 study by Swedish research that confirmed the occurrence in adult and old cats, too. Dragutin Lončar and Björn Afzelius at the Wenner Institute of the University of Stockholm showed that new-born kittens have fat resembling BAT subcutaneously (under the skin) and surrounding several internal organs (Lončar & Afzelius, 1989).

The structure of fat cells changes during a cat's development and aging. And simultaneously with the growth of the cat, the characteristics of fat cells change accordingly. The "brown fat" in kittens turns out to be "white fat" in adults. At the same time, the cell size/volume and spaces between cells also change into a structure resembling normal fat in adult mammals. The functions of cells also change, as the richness of veins typical for brown fat decreases and the number of mitochondria diminishes. Thus, during the ontogeny, the energy metabolism of cats significantly changes (Lončar & Afzelius, 1989).

Cold air activates brown fat and enhances resistance

Marked changes take place in a cat's fat tissues, both quantitatively and qualitatively as the animal is exposed to cold environmental conditions (cold air). Doctor Dragutin Lončar has studied for several years the cats' energy metabolism and occurrence of brown fat in the universities of Stockholm and Zagreb. When studying the effects of very cold air on the cat's fat metabolism, a set of young kittens (age 10–13 weeks) were kept for a week at extremely freezing temperatures, at minus 30 degrees centigrade (- 30 °C) twice a day, for seven days (Lončar *et al.*, 1986).

At the end of the experiment, the kittens were killed, and the amount, quality, and distribution of fat cells and tissues in cats' bodies were analyzed in detail by electronic microscopy. The fat tissues were taken for analyses from five different locations of the cats' body – directly beneath the skin, around the kidneys, around the heart, around the scapula (shoulder blade), and underarms. The cold exposure changed the properties of the cats' fat tissues in every target studied.

Cats in the control group were kept at normal house temperature for a week, The fat tissues in these cats were customary white fat throughout the study period, and the size of solitary cells and the cell cavities stayed at normal scale.

The size of fat cells in the cats exposed to cold diminished significantly. The size decreased by two-thirds at the maximum. The study also demonstrated changes in blood vessels in cold-exposed kittens. Due to the cold exposure, the number of capillaries in cell walls doubled during the week's experiment (Lončar *et al.*, 1986).

The most important result in the studies was the change in the number of mitochondria. The total volume of cats' mitochondria increased due to occasional cold exposures by fourfold. Detailed analyses of the inner structure of the "power plants of cells" confirmed that cold exposures significantly increased the number of mitochondria.

Repeated exposures to cold air led to changes in cats' fat tissues and metabolism that equal the changes and build-up of brown fat in other animal species, especially small rodents. At the time of those studies (the mid-1980s), the occurrence of brown fat in cats was not known in scientific literature.

The researchers concluded that the fat tissues in cats change the energy metabolism and that the changes enhance the cats' adaptation to a cold environment in any event where such a change is necessary (Lončar *et al.*, 1986).

Conclusions

The decision between the indoor or outdoor living for their cats divides cat owners' attitudes – and sometimes far too strictly. Denying outdoor access is mostly justified by various safety issues such as preventing traffic accidents or risks of infections, swallowing poisons, and in extreme situations, falling victim to predators or human cat-haters. In temperate and cold climate zones, the sensitivity to cold air is often argued as a reason to keep the moggy entirely indoors.

On the other hand, access to outdoor environments enhances any cat's well-being, both physical and mental. And the controversial question of cats' ability to withstand cold weather is largely misunderstood. Extreme attitudes arguing that cats cannot cope in cold are based on false assumptions. An indisputable fact is that the body temperature of cats is higher than that of men. This does not, however, mean that cats cannot go and roam outside.

The cats' cold adaptation and acclimatization are mostly studied in laboratories in various veterinary experiments. These have proved short-term tolerance to very cold conditions, even when the body temperature reaches dangerous hypothermic values. In everyday life, both domestic and feral or stray cats have proved the ability to acclimatize to harsh conditions during cold weather. Cats' natural ability to find shelter is well-developed but safe refuge should always be provided to domestic and, if possible, to feral and stray cats, too. And every cat needs necessarily unfrozen food and water. Advice to provide all the necessary services to outdoor cats is available to every cat owner and cat lower.

The most important basis for cats' cold tolerance is the fur. Most cats have thick fur, consisting of various hair types that provide efficient thermal insulation. The cats' fur is denser than that of dogs. A few cat breeds are bald or have very thin fur, and such moggies should not go outside in the cold.

Cats have also a significant anatomical feature ensuring resistance to cold. Cats have layers of brown adipose fat. i.e., tissues that can store and liberate warmth when necessary. And these tissues get stronger when the cats spend time in cold or cool conditions outside. The more a cat spends outdoors, the stronger the tolerance towards freezing conditions.

In a nutshell: Healthy adult cats tolerate cool and cold outdoor conditions, but any owner or friend of felines should know the animals' needs and restrictions and provide help when necessary.

References

Adams T. 1963. Mechanisms of cold acclimatization in the cat. *Journal of Applied Physiology,* 18(4): 778–780; https://doi.org/10.1152/jappl.1963.18.4.778

Animal Humane Society. (2022). Can outdoor cats survive Minnesota's sub-zero temps?; https://www.animalhumanesociety.org/news/can-outdoor-cats-survive-minnesotas-sub-zero-temps

Argos Pet Insurance. 2020, updated 2022. Everything you need to know about your cat's health this winter; https://www.argospetinsurance.co.uk/we-talk-pet/everything-you-need-to-know-about-your-cats-health-this-winter/

AVMA. American Veterinary Medical Association. 2022; Cold weather animal safety | American Veterinary Medical Association (avma.org)

Baker PJ, Soulsbury, CD, Iossa G & Harris S. 2010. Domestic cat (*Felis catus*) and Domestic dog (*Canis familiaris*). In: S.D. Gehrt, S.P.D. Riley & B.L. Cypher (Editors). *Urban Carnivores. Ecology, Conflict, and Conservation.* Baltimore: The Johns Hopkins University Press, pp. 157–171.

Bessant, C. (2006). *The Secret Life of Cats. Everything Your Cat Would Want You To Know.* 292 pp. London: John Blake Publishing Ltd.

Brodeus A, Wright A & Cortes Y. 2017. Hypothermia and targeted temperature management in cats and dogs. *Journal of Veterinary Emergency and Critical Care,* 27(2): 151–163; DOI: 10.1111/vec.12572

Brown S. 2020. *The Cat. A Natural History.* London: The Ivy Press.

Cannon B & Nedergaard J. 2004. Brown adipose tissue: function and physiological significance. *Physiological Reviews* 84(1): 277–359; DOI: 10.1152/physrev.00015.2003

Cats Protection. 2022. Cats and cold weather. Keep your cat safe in cold weather with our expert guide; Cats & cold weather | Help & Advice | Cats Protection

Clark MH, Ferguson DC, Bunick D & Hoenig M. 2013. Molecular and histological evidence of brown adipose tissue in adult cats.
The Veterinary Journal 195(1): 66–72; DOI: 10.1016/j.tvjl.2012.05.029

Coop Zeitung. 2011. Wenn das Büsi friert; Wenn das Büsi friert | Coopzeitung

Dozeman E & Dacvecc DVM. 2020. *MedVet,* 9.1.2020; Treatment of Hypothermia in Dogs and Cats - MedVet (medvetforpets.com)

Ein Herz für Tiere. (2020). Katzen im Winter. Die 12 wichtigsten Fakten; Katzen im Winter: Die 12 wichtigsten Fragen und Antworten | herz-fuer-tiere.de

Fogle B. 2003. *Cat Owner's Manual.* London: Dorling Kindersley Books.

Foreman-Worsley R, Finka LR, Ward SJ & Farnworth MJ. 2021. Indoors or outdoors? An international exploration of owner demographics and decision making associated with lifestyle of pet cats. *Animals,* 2021, 11, 253; https://doi.org/10.3390/ani11020253

Halls, V. (2010). *The Secret Life of Your Cat.* London: Hamlyn.

Health, S. (2001). *Cat & Kitten Behaviour. An Owner's Guide.* London: HarperCollinsPublishers.

Heidenberger E. 1997. Housing conditions and behavioural problems of indoor cats as assessed by their owners. *Applied Animal Behaviour Science,* 52 (3–4): 345–364; https://doi.org/10.1016/S0168-1591(96)01134-3

Hensel H & Banet M. 1982. Adaptive changes in cats after long-term exposure to various temperatures. *Journal of Applied Physiology,* 52(4): 1008–1012; https://doi.org/10.1152/jappl.1982.52.4.1008

Hill K. 2022. Brief description on evolution and habitat of wild cat. *Research & Reviews: Journal of Veterinary Sciences,* 6(3): eISSN:2581-3897 RRJVS; https://www.rroij.com/open-access/brief-description-on-evolution-and-habitat-of-wild-cat.pdf

Hilmer S, Algar D, Neck D & Schleucher E. 2010a. Remote sensing of physiological data: Impact of long-term captivity on body temperature variation of the feral cat (*Felis catus*) in Australia, recorded via Thermochron iButtons. *Journal of Thermal Biology* 35(5): 205–210, 2010; https://doi.org/10.1016/j.jtherbio.2010.05.002

Hilmer S, Algar D, Plath M & Schleucher E. 2010b. Relationship between daily body temperature and activity patterns of free-ranging feral cats in Australia. *Journal of Thermal Biology* 35(6): 270–274; https://doi.org/10.1016/j.jtherbio.2010.06.002

International Cat Care. 2020; Keeping Cats Safe: Anti-freeze | International Cat Care (icatcare.org)

Kays RW & DeWan AA. 2004. Ecological impact of inside/outside house cats around a suburban nature preserve. *Animal Conservation* 7(3): 273–283; https://doi.org/10.1017/S1367943004001489

Kays R, Dunn R. *et al.* (13 authors). 2020. The small home ranges and large local environmental impacts of pet cats. *Animal Conservation* 23(5): 516–523; https://doi.org/10.1111/acv.12563

Lieblingstier. 2022. Wann friert eine Katze?; Wann friert eine Katze? - Lieblingstier

Lončar D & Afzelius BA. 1989. Ontogenetical changes in adipose tissue of the cat: convertible adipose tissue. *Journal of Ultrastructure and Molecular Structure Research* 102(1): 9–23; https://doi.org/10.1016/0889-1605(89)90028-1

Lončar D, Bedrica L, Mayer J, Cannon B, Nedergaard J, Afzelius BA & Svaiger A. 1986. The effect of intermittent cold treatment on the adipose tissue of the cat. The apparent transformation from white to brown adipose tissue. *Journal of Ultrastructure and Molecular Structure Research* 97(1-3): 119–129; https://doi.org/10.1016/S0889-1605(86)80012-X

Loss SR, Boughton B, Cady SM, Londe DW, McKinney C, O'Connell T, Riggs GJ & Robertson EP. 2022. Review and synthesis of the global literature on domestic cat impacts on wildlife. *Journal of Animal Ecology* 91(7): 1361–1372; https://doi.org/10.1111/1365-2656.13745

Medivet. 2022. Caring for your cat in snow and freezing temperatures; Care for Cats & Kittens in the Cold Snowy Winter | Medivet

Mikkola S, Salonen M, Hakanen E & Lohi H. 2022. Fearfulness associates with problematic behaviors and poor socialization in cats. *iScience* 25(10): 105265; https://doi.org/10.1016/j.isci.2022.105265

Milne E & Wild K. 2018. *The Purrfect Guide to Thinking Like a Cat. 501 Tips and Techniques.* London: Amber Books Ltd.

Morris D. 1994. *Catwatching.* London: Random House.

Nadine. 2020. Ab wann frieren Katzen? 6 Tipps zum Erkennen, Wärmen, Vorbeugen. Das Katzenmagazin - *CatNews*; Ab wann frieren Katzen? 6 Tipps zum Erkennen, Wärmen, Vorbeugen | Das Katzenmagazin (cat-news.net)

Olmstead Ch E, Villablanca JR, Torbiner M & Rhodes D. 1979. Development of thermoregulation in the kitten. *Physiology & Behavior* 23(3): 489–495; https://doi.org/10.1016/0031-9384(79)90048-9

Parker M, Serra J, Deputte BL, Ract-Madoux B, Faustin M & Challet E. 2022. Comparison of locomotor and feeding rhythms between indoor and outdoor cats living in captivity. *Animals*, 12(18): 2440; https://doi.org/10.3390/ani12182440

Peiris AN, Jaroudi S & Gavin M. (2018). Hypothermia. *JAMA*. 319(12): 1290; https://jamanetwork.com/journals/jama/fullarticle/2676112

PetMD, Editorial. 2009. Low Body Temperature in Cats; Low Body Temperature in Cats | PetMD

Piamore E. 2019. Cat History and Evolution. *AnimalWised*, 5.9.2019; https://www.animalwised.com/cat-history-and-evolution-3277.html

Piccione G, Marafioti S, Giannetto C, Panzera M & Fazio F. 2013. Daily rhythm of total activity pattern in domestic cats (*Felis silvestris catus*) maintained in two different housing conditions. *Journal of Veterinary Behavior Clinical Applications and Research* 8(4): 189–194; https://doi.org/10.1016/j.jveb.2012.09.004

Randall W, Cunningham JT, Randall S, Luttschwager J & Johnson RF. 1987. A two-peak circadian system in body temperature and activity in the domestic cat, *Felis catus* L.; *Journal of Thermal Biology* 12(1): 27–37; https://doi.org/10.1016/0306-4565(87)90020-9

Steen PA, Soule EH & Michenfelder JD. 1979. The detrimental effect of prolonged hypothermia in cats and monkeys with and without regional cerebral ischemia. *Stroke* 10(5): 522–529; https://doi.org/10.1161/01.STR.10.5.522

Stella JL & Croney CC. 2016. Environmental aspects of domestic cat care and management: Implications for cat welfare. *The Scientific World Journal*, Volume 2016; Article ID 6296315; https://doi.org/10.1155/2016/6296315

Svedea. 2022. Kan utekatter vara ute på vintern? [Can outdoor cats spend time in outdoors in Winter? (In Swedish)]; Kan utekatter vara ute på vintern? (svedea.se)

Tan SML, Stellato AC & Niel L. 2020. Uncontrolled outdoor access for cats: An assessment of risks and benefits. *Animals* 10(2): 258; https://doi.org/10.3390/ani10020258

Tractive. 2022. Outdoor cat care in winter: Tips to keep your feline friends warm and safe; Outdoor Cat Care in Winter: Cold Weather Cat Safety Tips - Tractive

Uelzener Magazin. 2018. Frieren Katzen bei Kälte?; Frieren Katzen bei Kälte? | Uelzener

Veterinären.nu. 2022. Vilken temperatur klarar en kat? [What is the temperature cat can tolerate? (In Swedish).]; Veterinären.nu – Vilken temperatur klarar en katt? (veterinaren.nu)

Wagner D, Hurley K & Stavisky J. 2018. Shelter housing for cats: Practical aspects of design and construction, and adaptation of existing accommodation. *Journal of Feline Medicine and Surgery*, 20(7): 643–652; https://doi.org/10.1177/1098612X18781390

Zhang Z, Li Y, Ullah S, Chen L, Ning S, Lu L, Lin W. & Li Z. 2022. Case Report Home Range and Activity Patterns of Free-Ranging Cats: A Case Study from a Chinese University Campus. *Animals* 12(9): 1141; https://doi.org/10.3390/ani12091141

YOUR KNOWLEDGE HAS VALUE

- We will publish your bachelor's and
  master's thesis, essays and papers

- Your own eBook and book -
  sold worldwide in all relevant shops

- Earn money with each sale

Upload your text at www.GRIN.com
and publish for free